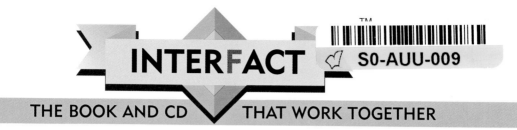

INTERFACT

THE BOOK AND CD ⬥ THAT WORK TOGETHER

SPACE TRAVEL

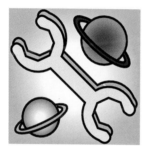

TWO CAN ™

LONDON • CHANHASSEN, MINNESOTA

Copyright © 2002, 1999 by Two-Can Publishing

Two-Can Publishing
An imprint of Creative Publishing international, Inc.
15 New Bridge Street
London EC4V 6AU
www.two-canpublishing.com

Created by
act-two
346 Old Street
London
EC1V 9RB

'Two-Can' and 'Interfact' are trademarks of Creative Publishing international, Inc.
18705 Lake Drive East, Chanhassen, Minnesota 55317 USA

ISBN 1-85434-917-1

A catalogue record for this book is available from the British Library.

Photographic Credits: Front cover: NASA/Science Photo Library; p.8 Science Photo Library;
p.9 European Space Agency/Ian Graham; p.10 Associated Press; p.11 NASA;
p.13 NASA/Spacecharts; p.14 NASA; p.15 NASA; p. 16 NASA; p.17 NASA/Spacecharts;
p.19 NASA; p.20 NASA; p.21 NASA/Spacecharts; p.22 NASA/Science Photo Library; p.23 (top)
NASA, (bottom) Science Photo Library; p.24 US Naval Observatory/Science Photo Library; p.25 (top)
NASA, (bottom) NASA/Science Photo Library; p.27 NASA; p.36 NASA; p.37 NASA/Spacecharts

Every effort has been made to acknowledge correctly and contact the source
of each picture, and Two-Can apologises for any unintentional
errors or omissions, which will be corrected in future editions of this book.

2 3 4 5 6 08 07 06 05 04 03

Printed in Hong Kong

INTERFACT

THE BOOK AND CD THAT WORK TOGETHER

INTERFACT will have you hooked in minutes – and that's a fact!

⬤ **The disk is packed with interactive games, puzzles, quizzes and activities that are challenging, fun, and full of interesting facts.**

Meet the astronauts who live on a space station and find out how they live.

Explore the screen with your mouse

⬤ **Open the book and discover more fascinating information highlighted with lots of full-colour illustrations and photographs.**

Satellites

The **Soviet Union** launched the world's first artificial **satellite**, Sputnik 1, in 1957. It travelled around the Earth and transmitted radio signals back to scientists in the Soviet Union. Today, there are over 1,000 satellites operated by more than 20 countries in orbit around the Earth. Satellites affect our lives in many ways and have made the world feel like a much smaller place. Communications satellites transmit live television pictures from the other side of the world and connect us by telephone to people many kilometres away. Weather satellites provide up-to-the-minute information on the weather situation anywhere in the world, while navigation satellites help pilots and sailors plot a course across the sky or ocean.

▶ This photograph shows a communications satellite soaring high above the Earth. The large "wings" on either side of the satellite are solar panels. These obtain energy from the Sun, which is then used to power the satellite. A communications satellite receives information from a building on Earth called a ground station. It then transmits this information to a ground station somewhere else in the world.

DISK LINK
Set up your own satellite launch business in ACME SATELLITES.

◀ Thanks to weather satellite images like this, forecasters can warn people of severe weather conditions before they strike. This large white spiral shows a hurricane moving across the southern United States. The land shows up as green in this image.

...e slightly blurred atmosphere, and as a result, escapes its distorting effect.

◀ The Hubble Space Telescope seen against the blackness of space.

Read and learn about satellites and their many uses.

⬤ To get the most out of **INTERFACT**, use the book and disk together. Look out for the special signs, called Disk Links and Bookmarks. To find out more, turn to page 41.

41

BOOKMARK

DISK LINK
We have lift off! You'll soon rocket up the board when you play the SPACE RACE.

Once you've clicked on to **INTERFACT**, you'll never look back.

LOAD UP!
Go to **page 38** to find out how to load your disk and click into action.

HELP SCREEN

Learn how to use the disk in no time at all.

Welcome to the

INTERFACT

disk on Space Travel

To have a look at all the different things on the disk, simply click the arrow keys with your mouse.

As you do this, you'll see a description of each activity in the reading box.

Click on the picture at the top of the screen to select the activity you want to investigate.

Get to grips with the controls and find out how to use:

- arrow keys
- reading boxes
- 'hot' words

THE SPACE RACE

The space-themed board game that won't leave you bored!

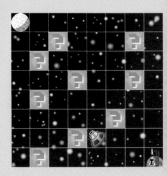

This rocket-powered board game is 'snakes and ladders' with a difference. Land on a rocket square and you'll blast up the board, but land on a capsule and you'll float back down again!

MOON QUEST

Blast off in our challenging moon landing quiz!

Every correct answer in this exciting quiz will edge your spacecraft towards the Moon. But watch out – an incorrect answer will cost you valuable fuel!

ACME SATELLITES

Make your fortune putting satellites into orbit!

Can you build up a global satellite launch business? It'll take stacks of skill and satellite know-how to make each mission a success.

THE NAME GAME

A puzzling who's who of the history of space travel.

These famous space travellers all look the same with their spacesuits on! The only way you can work out who's who is to ask each one of them about their travels in space.

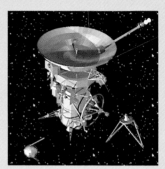

REPAIR SHOP

Have you got what it takes to become a space mechanic?

If you've never patched up a space probe, then here's your chance! The Cassini-Huygens space probe has been damaged during its mission, and it's up to you to repair it.

FLOATING FREE

Learn about daily life on board a space station.

Meet the astronauts who spend many months on board a space station. Find out what they like best and what they hate most about life in space.

PAST PROBES

Chart the progress of space probes through history.

In this game, it's your job to monitor space probes. Travel through time to check on their progress as they cross the Solar System.

What's in the book

*All words in the text that appear in **bold** can be found in the glossary*

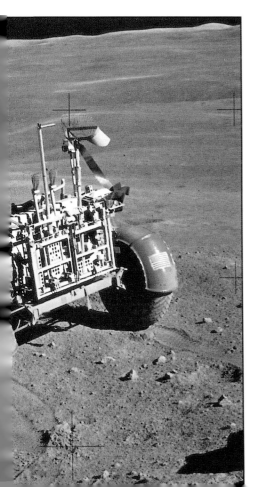

Rockets

People have always dreamed of travelling into space, but it was not until the mid-1900s that the dream became possible. The invention of the **liquid fuel** rocket in the 1920s was the first major step on the way to space travel. However, it was only during the 1950s that the technology for space rockets was fully developed.

DISK LINK
We have lift off! You'll soon rocket up the board when you play the SPACE RACE.

A spacecraft needs to travel at great speed to reach space. One of the few vehicles powerful enough to carry a spacecraft into **orbit** is a liquid fuel rocket. Its engine works by burning huge amounts of liquid fuel. The fuel is pumped into a burning chamber, where it is mixed with liquid oxygen. Fuel needs oxygen to burn, and because there is no oxygen in space, the rocket must carry its own supply.

Balloon

Air

▲ Robert H Goddard standing beside the first successful liquid-fuel rocket in 1926.

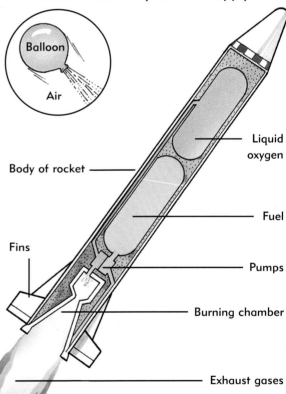

Body of rocket

Fins

Liquid oxygen

Fuel

Pumps

Burning chamber

Exhaust gases

▲ Burning gases rush out of the rocket, pushing it in the opposite direction like a balloon with an open neck.

▲ This picture shows the Titan 3E-Centaur rocket launching the unmanned Voyager 1 **space probe** in 1977. Voyager 1 has explored the **planets** Jupiter and Saturn, and continues to travel through space today.

ROCKET FACTS

● Chinese soldiers used an early type of rocket in the 13th century AD. It was powered by gunpowder and used as a weapon against attacking armies.

● Some rocket engines are more than 3,000 times more powerful than a car engine of the same size.

● Rockets are steered by swivelling their nozzles to direct the burning gases in different directions.

● The inventor of the liquid-fuel rocket, Robert H Goddard, had to stop testing rockets in his home town because of complaints about the noise. He continued his experiments in an isolated spot in the desert, where he could make as much noise as he liked!

● The fastest ever rocket-propelled object is the Helios B space probe, which was sent to investigate the Sun. As it orbits the Sun, Helios B sometimes reaches speeds of up to 252,800kph.

● The temperature in some rocket engines reaches 3,300°C. That's about twice the temperature at which steel melts!

Early spaceflights

The space age began on October 4, 1957, when the **Soviet Union** launched a rocket carrying Sputnik 1, the world's first **satellite**. A month later, the Soviets launched a second satellite, called Sputnik 2. It carried a dog called Laika (meaning 'barker'), who became the first living creature in space.

In 1961, the Soviet Union sent the first man, Yuri Gagarin, into space. His Vostok 1 spacecraft made one **orbit** of the Earth before returning home safely. The following year, John Glenn became the first American to orbit the Earth. He travelled around the Earth three times on board the Mercury space capsule.

DISK LINK
Meet some of the earliest astronauts when you play the NAME GAME.

▲ **Astronaut** John Glenn prepares for the United States' first manned orbital spaceflight in 1962. Glenn took five hours to orbit the Earth three times.

▶ The first joint mission between the Soviet Union and the United States took place in 1972. Shown here are Soviet **cosmonaut** Valery Kubasov and US astronaut Thomas Stafford.

◀ The world's first space traveller, Soviet cosmonaut Yuri Gagarin, waits to board his Vostok 1 spacecraft.

Saturn V rocket

The success of the early space flights inspired the US to plan their most ambitious space travel project yet – a mission to the Moon with **astronauts**. The Saturn V rocket, which was built to carry people to the Moon, was at the time the largest, most powerful rocket ever built by the United States.

Saturn V consisted of three sections, called stages. The spacecraft, which carried the astronauts, was attached to the top of the rocket. Each stage carried a huge quantity of **liquid fuel**. When this fuel ran out, the stage was no longer needed and dropped safely away.

▶ This illustration shows a cross-section of the Saturn V rocket. Over seven years, Saturn V rockets were used to launch seven Apollo missions and put a Skylab **space station** into space.

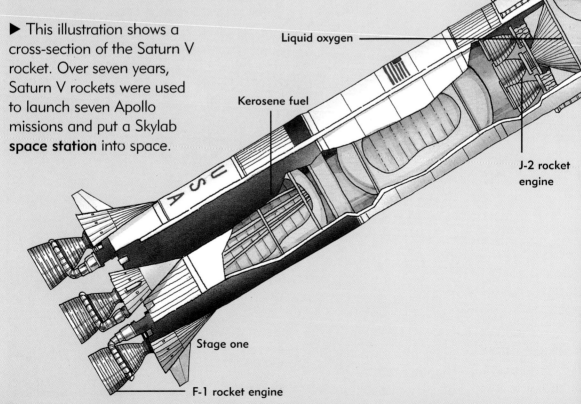

Stage two

Liquid oxygen

Kerosene fuel

J-2 rocket engine

Stage one

F-1 rocket engine

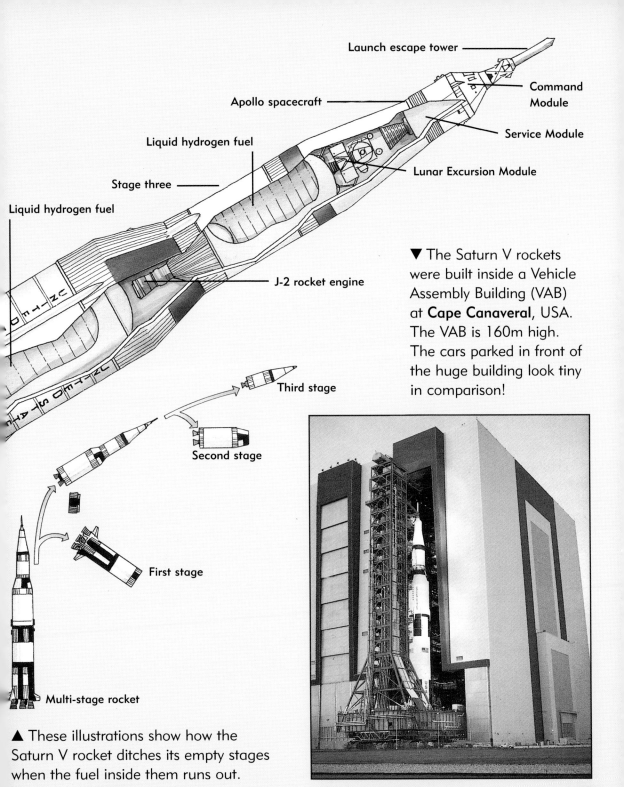

Launch escape tower

Command Module

Apollo spacecraft

Service Module

Liquid hydrogen fuel

Lunar Excursion Module

Stage three

Liquid hydrogen fuel

J-2 rocket engine

Third stage

Second stage

First stage

Multi-stage rocket

▲ These illustrations show how the Saturn V rocket ditches its empty stages when the fuel inside them runs out.

▼ The Saturn V rockets were built inside a Vehicle Assembly Building (VAB) at **Cape Canaveral**, USA. The VAB is 160m high. The cars parked in front of the huge building look tiny in comparison!

The Apollo Project

The United States called its plan to put **astronauts** on the Moon 'the Apollo Project'. They spent years developing the Apollo spacecraft. It consisted of three parts: the Command Module, the Service Module and the Lunar Module. The Command Module carried the three

▼ A Saturn V rocket ready for take-off on its launch pad at **Cape Canaveral**.

astronauts from the launch pad into **orbit** around the Moon. Two of the astronauts would then climb into the Lunar Module and land on the Moon, while the third astronaut stayed on board the Command Module. Later, the Lunar Module would return to the Command Module, in which all three astronauts would return home. During the mission, the Service Module would provide air, water and power.

▼ Two Apollo astronauts would land on the Moon's surface in the Lunar Module. One astronaut stayed behind in the Command Module in orbit around the Moon. The sequence below shows how the Apollo astronauts took off from the Moon and returned to Earth.

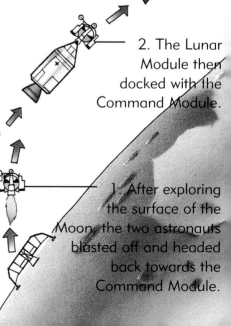

2. The Lunar Module then docked with the Command Module.

1. After exploring the surface of the Moon, the two astronauts blasted off and headed back towards the Command Module.

MOON FACTS

● The first people to travel beyond the Earth's orbit were the three crew members of Apollo 8, who circled the Moon in 1968. They were the first people to see 'the dark side of the Moon'.

● More than a million spectators stood at Cape Canaveral to watch the launch of the Apollo 11 spacecraft.

4. The Command Module's booster rockets then fired, carrying the astronauts back towards Earth.

3. The two astronauts returned to the Command Module. The empty Lunar Module, which was no longer needed, was ditched.

▼ 5. At the end of the journey, the Command Module landed in the sea. An inflatable collar and bags ensured that it stayed afloat until the astronauts could be transferred to a nearby recovery ship.

Man on the Moon

On July 21, 1969, Apollo 11 **astronauts** Neil Armstrong and Edwin 'Buzz' Aldrin stepped on to the Moon. A third astronaut, Michael Collins, stayed in the Command Module during the historic landing.

During their short stay on the Moon, the Apollo astronauts set up experiments, took photographs and collected rock samples. Their back-packs supplied air and kept them cool by pumping cold water through tubes in their spacesuits.

Over the next three years, ten more astronauts would return to the Moon as part of the Apollo Project.

▲ Apollo 11 astronaut Edwin 'Buzz' Aldrin steps down from the Lunar Module. He was the second man on the Moon, after Neil Armstrong. The two astronauts set up experiments and collected rocks.

◄ The Apollo 11, 12 and 14 crews moved around the Moon on foot. They could walk up to 4km from the Lunar Module. Apollos 15, 16 and 17 carried a battery-powered car called a Lunar Rover, which enabled them to travel much greater distances. The Apollo 15 astronauts spent more than 18 hours driving around the Moon's surface.

DISK LINK
Blast off in a Saturn V rocket when you play
MOON QUEST.

DID YOU KNOW?

● The Lunar Rover weighed 209kg and could travel at up to 16kph. Three Lunar Rovers were left behind on the Moon by Apollo missions.

● The Apollo 14 astronauts Alan Shepard and Edgar Mitchell spent nine hours on the Moon. Before returning to the Command Module, Shepard played golf!

● The six Apollo crews that landed on the Moon brought a total of 379kg of Moon rock and soil back to Earth.

● The footprints made on the Moon by the Apollo astronauts will stay there forever because there is no wind or rain on the Moon to blow or wash them away.

Space Shuttle

Launching rockets such as the Saturn V is incredibly expensive because the rockets can only be used once. During the late 1970s, **NASA** scientists began working on the Space Shuttle, a spacecraft that could be used again and again. This would greatly reduce the cost of space travel.

The part of the Shuttle that carries the **astronauts** – the orbiter – is launched by three rockets. Once each of these is empty, the orbiter falls away. At the end of its mission, the orbiter re-enters the Earth's **atmosphere**. It glides down and lands on a runway, rather like a jet aircraft.

▶ The orbiter contains a huge cargo hold called the **payload** bay. Here, the orbiter's robot arm is being used to move a **satellite** out of the bay and place it into **orbit**.

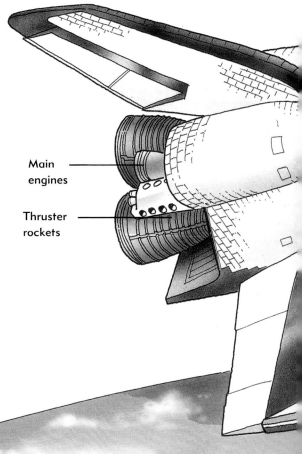

Main engines

Thruster rockets

SHUTTLE FACTS

● The orbiter has 44 small rocket engines in its nose and tail to adjust the Shuttle's position.

● During pre-launch tests, the Shuttle was carried on the back of a jumbo jet. While the jet was flying, the Shuttle was released to test its ability to glide back to Earth and land safely.

● The first Space Shuttle was called the *Enterprise*. It was named after the spaceship in the TV series *Star Trek*!

● During **re-entry**, astronauts on board the Shuttle cannot speak to Earth by radio. This is called a 'radio blackout'.

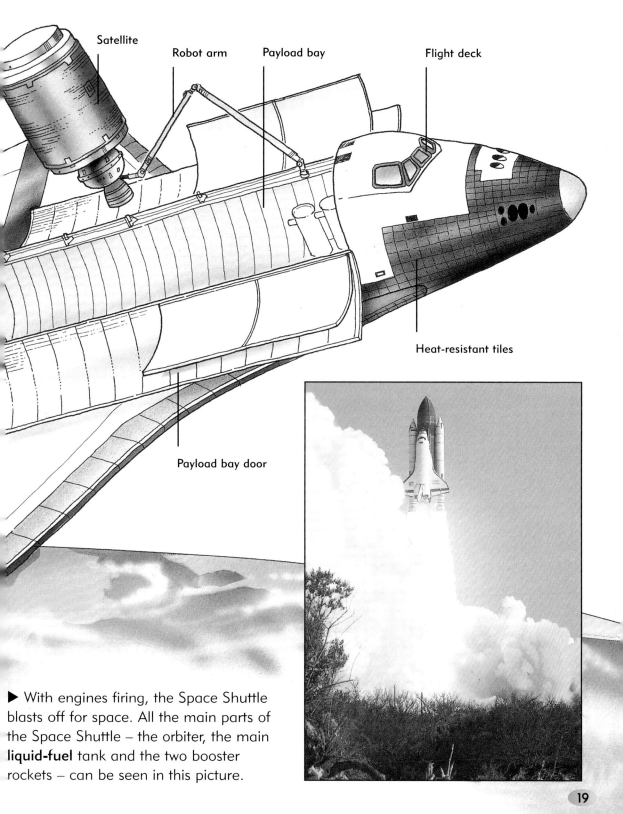

Satellite

Robot arm

Payload bay

Flight deck

Heat-resistant tiles

Payload bay door

▶ With engines firing, the Space Shuttle blasts off for space. All the main parts of the Space Shuttle – the orbiter, the main **liquid-fuel** tank and the two booster rockets – can be seen in this picture.

Shuttle missions

The Space Shuttle was designed mainly as a 'space truck' to carry **satellites** into **orbit**. Engineers from the Space Shuttle can also repair satellites that are already in orbit, or bring them back to Earth if they go wrong.

Up to seven **astronauts** can ride in the Space Shuttle – many more than on earlier spacecraft. Only Space Shuttle pilots need to be trained astronauts, so the other crew members may include doctors, engineers or biologists, who wish to attempt experiments that can only be carried out in space. Shuttle missions last for up to 30 days.

DID YOU KNOW?

● During **re-entry**, air rushing past the orbiter heats it to over 1,300°C. Thousands of special tiles glued onto the underside of the orbiter act as a vital heat shield.

▼ Its mission completed, the orbiter lands at over 300kph on the dry lake bed at the Edwards Air Force Base in California.

▶ Astronauts check equipment inside the orbiter **payload** bay. A tether, or lifeline, prevents them from floating away.

Satellites

The **Soviet Union** launched the world's first artificial **satellite**, Sputnik 1, in 1957. The satellite travelled around the Earth and transmitted radio signals back to scientists in the Soviet Union. Today, there are over 1,000 satellites, operated by more than 20 countries, in orbit around the Earth.

Satellites affect our lives in many ways and have made the world feel like a much smaller place. Communications satellites transmit live television pictures from the other side of the world and connect us by telephone to people many kilometres away. Weather satellites provide up-to-the-minute information on the weather situation anywhere in the world, while navigation satellites help pilots and sailors plot a course across the sky or ocean.

The Hubble Space Telescope was launched in 1990 to collect images and take measurements from distant stars and galaxies. Satellite telescopes, such as Hubble, produce much clearer images than telescopes on the ground. The Earth's **atmosphere** bends light from stars and galaxies, and so images from telescopes on Earth are slightly blurred. Hubble orbits well above the Earth's atmosphere and, as a result, escapes its distorting effect.

◀ The Hubble Space Telescope seen against the blackness of space.

▶ This photograph shows a communications satellite soaring high above the Earth. The large 'wings' on either side of the satellite are solar panels. These obtain energy from the Sun, which is then used to power the satellite. A communications satellite receives information from a building on Earth called a ground station. It then transmits this information to a ground station somewhere else in the world.

DISK LINK
Set up your own satellite launch business in ACME SATELLITES.

◀ Thanks to weather satellite images like this, forecasters can warn people of severe weather conditions before they strike. The large white spiral in this picture shows a hurricane moving across the southern USA. The land shows up as green in this image.

Space probes

A space probe is a vehicle with no crew launched to explore space. Some probes are sent to **orbit** a **planet** or moon, and some are sent to land on a planet to collect samples and take measurements. Others are sent to investigate comets, or even fly around the Sun. To date, space probes have visited every planet in our Solar System apart from Pluto.

A probe's mission may last for many years. The Voyager 2 space probe, which was launched in 1977, has flown past and photographed the planets Jupiter, Saturn, Uranus and Neptune. It took 12 years for Voyager to reach Neptune and, today, it continues to travel through the outer reaches of the Solar System.

▼ The Galileo space probe was launched by the United States in 1989. It reached the planet Jupiter in 1995 and spent two years orbiting the giant planet. Galileo consisted of two parts: an atmosphere probe and an orbiting spacecraft.

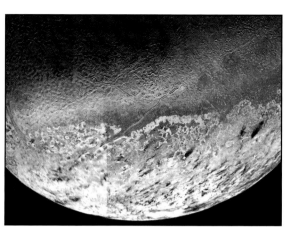

DISK LINK

Take a look at space probes through history in PAST PROBES.

▲ This image of Triton, a moon of the planet Jupiter, was taken by the Voyager 2 space probe in 1989. One of Voyager's many discoveries was that Triton is covered with geysers – cracks in the ground through which hot liquid and gases erupt.

▼ In July 1997, the Mars Pathfinder space probe landed on the planet Mars. Its mission was to study the planet's surface and to assess the possibility of further explorations. The Pathfinder carried a mini-rover, called the Sojourner, which can be seen in the middle of the photograph below. The Sojourner, which was controlled by scientists on Earth, roamed the planet for three months, collecting images and taking measurements. Contact with the rover ended in September 1997.

Space station

A space station is a large spacecraft designed to stay permanently in **orbit**. Crews live and work in space stations for several months at a time. The **Soviet Union** launched the first space station, which was called Salyut 1, in 1971.

Two years later, the United States launched the Skylab space station.

In 1986, the Soviets launched a larger space station, called Mir. Since 1986, Mir has been an important site for space research. Some **astronauts** have lived and worked on Mir for more than a year.

DISK LINK
Want to find out about life on a space station? Then check out FLOATING FREE.

▼ A new international space station has been planned for many years, but remains unbuilt. This artist's impression shows a Space Shuttle bringing supplies to the planned space station.

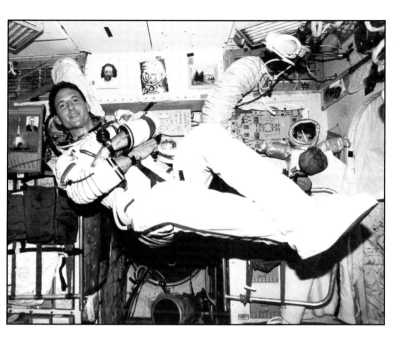

◀ This photograph shows the United States astronaut Jerry Linenger on board Mir. Note the picture on the wall behind him showing Yuri Gagarin – a hero to astronauts and **cosmonauts** alike.

▼ Astronauts have plenty of time to relax on board Mir space station. Here, United States astronaut Shannon Lucid chooses a book to read from the space station's library.

The space students

How would you like to take part in a mission on board the Space Shuttle? The two young people in this story have dreamed of travelling into space for years, and as we join the story, they are waiting to see if they have been accepted for astronaut training school. If they have, their dream may soon become a reality!

Every day for the past week, John had been getting up earlier than anyone else in his family. Not because he was desperate to get to school, or because he wanted to bring his parents a cup of tea in bed. No, John had much more important things on his mind. Any day now, he expected the most important letter in his life to drop through the letterbox – the letter that would tell him whether or not he had been accepted for astronaut training school.

As John reached the bottom of the stairs, he saw that the post had arrived early. Even before he picked up the envelope, John could see the letters 'NASA' in bright red across the top. With shaking hands, he tore the envelope open. John stared at the letter for a second, and then grinned from ear to ear. He had been accepted!

Several streets away, Carol was going through the same agonising wait. She was sixteen – the same age as her friend John. Sitting in the kitchen, she heard a letter drop onto the doormat. It was from NASA. Too nervous to open it, she handed it to her mother to open. Her mother stared at the letter, turned to Carol and said: "You've done it!"

Arriving at the Johnson Space Center in Texas, where astronauts for every United States space mission are trained, John and Carol found it hard to contain their excitement. The first weeks of training would be spent in the classroom. They would learn how the Shuttle worked, what each piece of equipment did and how to stay healthy in space.

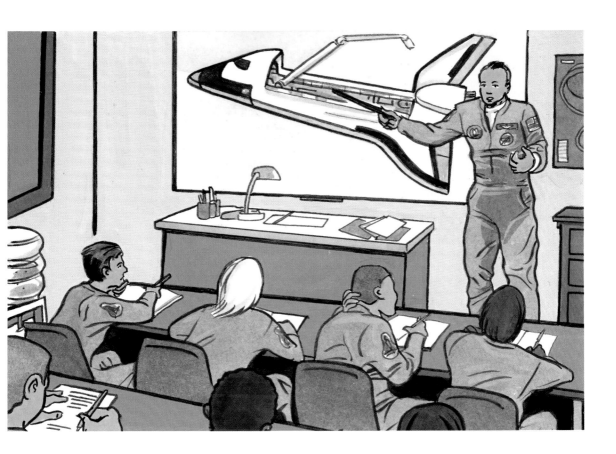

Some well-known astronauts visited the classroom and told the students about their experiences. One famous astronaut admitted that he had been very nervous about his first mission. He assured the class that this was not unusual, and to remember that nowadays space travel was very safe. It was amazing – much better than being at school!

At the end of the classroom training, the student astronauts prepared for the second part of their astronaut training – weightlessness. In space, astronauts float freely inside the Shuttle, so students must become used to carrying out their duties and performing everyday tasks on board when they can't tell up from down!

Astronauts train for weightlessness by flying in a special plane. The pilot flies the plane on a steep climb and then steers it into a dive. As the plane reaches the top of its climb and begins its descent, the passengers experience a feeling very similar to weightlessness.

Many students look forward to this part of the training more than any other, and John and Carol were no exception. Even if they never made it into space, at least they would be able to say that they had experienced a feeling similar to weightlessness. Not many people can say the same thing!

The inside of the plane consisted of little more than a large hollow tube with padded sides. As it climbed into the sky, John noticed nothing out of the ordinary, but as the plane began to fly through a series of arcing climbs and dives, he watched his feet lift off the ground, then stared in wonder as he watched Carol's whole body lift clear off the floor. This was excellent fun!

The following week, it was time for survival training. "Now," said the commanding officer, "it is very unlikely that anything will go wrong during a space mission, but if it does, you need to be prepared. For instance, if you crash land in an area that's difficult to reach, it may be several days before anyone is able to rescue you. During the next part of your training, you will learn how to survive just about any situation."

Parachuting in to a wilderness in northern Canada, all the students, including John and Carol, were expected to find their own food and water, construct their own shelters and wait for a rescue helicopter. Both John and Carol agreed that the three days of survival training were the hardest part of all. Cold, hungry and lonely, they both craved a hot meal and a warm bed. When they finally heard the rescue

helicopter approach, they leapt up and down, hugging each other with relief.

At the end of survival training, some of the students decided that the training was too difficult and dropped out. John and Carol, however, were determined to continue – they had come this far and refused to give up their chance of a place on the Shuttle. The instructor was pleased that they had decided to continue. He assured them that the hardest part was over. The remaining weeks of training would be spent on board the simulator, learning about the Shuttle's controls and getting to grips with the various space flight systems.

The earlier classroom training had prepared them well for their time in the simulator. They were amazed at how much they had learned – but daunted by how much there was still left to learn. The students had to learn how to use all kinds of flight controls, oxygen and power controls, emergency procedures, dials, switches and digital displays, and know exactly when to use them.

Learning all about the Shuttle's controls had seemed an impossible task, but at the end of 12 weeks' intensive simulator training, both John and Carol could move around confidently inside the cockpit, easily finding all the Shuttle's controls. What really mattered, though, was what the instructors thought of their abilities. Only they could award a pass or a fail.

Finally, the day came when the students would find out whether they had passed their training. One by one they were called into the head instructor's office. Some of the students came out looking delighted, while others emerged with their heads down, looking upset.

Soon it was John and Carol's turn. The walk to the instructor's office felt like the longest journey they had ever made. Within minutes, they would know whether or not they had been accepted as Shuttle astronauts. Carol knocked gingerly on the door and heard the instructor shout: "Come in!" They both stepped inside.

"Well, John and Carol," began the instructor, "you are by far the youngest students we have ever accepted in this school. And soon, thanks to your hard work, you will become the youngest ever members of a Shuttle team. You're going to space – congratulations!"

Build a space station

You can hang your model space station from the ceiling by a piece of string on a drawing pin.

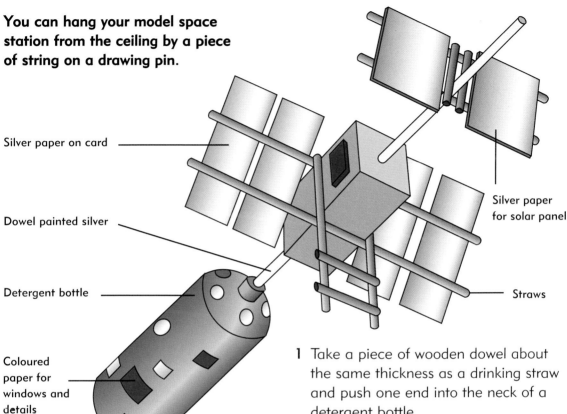

Silver paper on card

Dowel painted silver

Detergent bottle

Coloured paper for windows and details

Silver paper for solar panel

Straws

You can make your model space station from a few household items such as straws and cardboard.

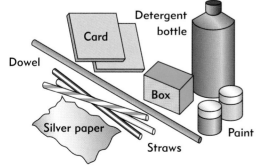

Card

Detergent bottle

Dowel

Box

Silver paper

Paint

Straws

1 Take a piece of wooden dowel about the same thickness as a drinking straw and push one end into the neck of a detergent bottle.

2 Push the other end of the dowel through a small cardboard box.

3 Now glue the drinking straws to the box and dowel as shown to form a framework to carry the solar panels.

4 Cut the solar panels out of cardboard sheets and cover them with silver paper.

5 Glue them to the straws.

6 Cut out windows from pieces of coloured paper and stick them to the bottle.

7 To add the finishing touches to your space station, paint it with model paints.

True or false?

Which of these facts are true and which are false? If you have read this book carefully, you will know the answers!

1. The first man in space was United States astronaut John Glenn.

2. Mercury spacecraft carried the first United States astronauts into space.

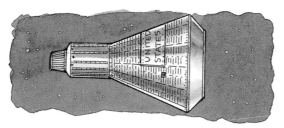

3. The Saturn V rocket was built to launch Soviet space stations.

4. There are about 100 satellites in orbit around the Earth.

5. Each successful Apollo mission landed two astronauts on the Moon.

6. The Apollo 14 astronauts drove around the Moon in a Lunar Rover.

7. The Space Shuttle was designed to carry satellites into orbit.

8. United States astronaut Alan Shepard played golf on the Moon during the Apollo 14 mission.

9. In 1965, Alexei Leonov became the first person to go on a spacewalk.

10. To date, space probes have visited every planet in the Solar System, apart from Jupiter.

11. Mir is the name of a Soviet space station.

12. The Voyager 2 space probe carried two astronauts to the planet Jupiter.

ANSWERS: 1.F 2.T 3.F 4.F 5.T 6.F 7.T 8.T 9.T 10.F 11.T 12.F

Lost in space!

The words in this puzzle are all lost in space, so photocopy this page and see if you can find them. The words may be written forwards, backwards or even diagonally. When you find a space-related word, cross it off from the list below.

V	O	Y	A	G	E	R	R	A	T	S	D
C	O	U	N	T	D	O	W	N	Y	B	O
O	E	L	G	A	E	P	P	M	T	A	C
R	K	C	O	L	R	I	A	E	I	L	K
B	R	T	I	U	S	R	Y	T	V	Y	O
I	E	S	H	U	T	T	L	E	A	K	E
T	T	A	S	S	A	M	O	O	R	S	L
S	S	T	U	N	G	A	A	R	G	U	E
M	O	U	N	U	E	E	D	G	N	L	B
A	O	R	O	B	O	T	B	A	S	E	O
R	B	N	N	O	O	M	R	S	P	U	R
S	E	T	I	L	L	E	T	A	S	F	P

airlock	eagle	mars	payload	shuttle	team
base	fuel	mass	probe	skylab	voyager
booster	gas	meteor	robot	stage	
countdown	gravity	moon	satellites	star	
dock	lunar	orbits	saturn	sun	

Glossary

Astronauts are people trained to travel into space. Astronaut means star traveller.

Atmosphere is the layer of gas surrounding a planet. The Earth's atmosphere consists mostly of nitrogen, but also contains oxygen.

Cape Canaveral is the place on the Florida coast where most United States rockets are launched.

Cosmonaut is a space traveller from Russia or the former Soviet Union (see Soviet Union). The first cosmonaut, and the first man in space, was Yuri Gagarin.

Gravity is the force that attracts objects towards each other. To reach space, rockets must break free of Earth's gravity.

Liquid fuels, such as kerosene or liquid hydrogen, are used to power most rockets.

NASA is the United States government agency that conducts space travel and research. It stands for the National Aeronautics and Space Administration.

Orbit is the curved path followed by an object circling a planet or star.

Payload is the name for the cargo carried by a rocket or Space Shuttle.

▲ The Space Shuttle Challenger, with its payload bay doors open, circles the Earth high above the atmosphere.

Planets are objects that revolve around a star. The planets in our Solar System are Mercury, Venus, Earth, Mars, Jupiter, Saturn, Uranus, Neptune and Pluto.

Re-entry describes the return of a spacecraft through Earth's atmosphere. As a spacecraft re-enters, air rushing past the outside heats it to high temperatures.

Satellite is an object that revolves around a planet or moon. It might be a natural satellite, such as the Earth's Moon, or a man-made satellite, like Sputnik 1 or the Hubble Space Telescope.

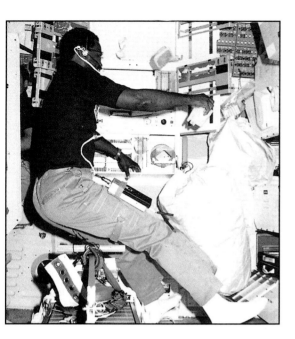

▲ Astronaut Ronald E McNair, weightless in space, prepares a meal for his fellow crew members on board the Shuttle.

Soviet Union was a huge country in central Asia that put the first satellite and the first man into space. In 1989, the Soviet Union broke up into many smaller countries, including Russia.

Space probe is a vehicle with no crew that is sent to explore space. Space probes have visited every planet in our Solar System, except Pluto, and have even been sent to orbit the Sun.

Space Station is a large structure permanently in orbit around the Earth, in which people live and work.

SPACE TIMECHART

1957 The Soviet Union launches the first manufactured object to reach space – Sputnik 1.

1961 Soviet Yuri Gagarin becomes the first person in space.

1962 John Glenn becomes the first United States astronaut to orbit Earth.

1962 The first live satellite TV pictures are relayed across the Atlantic Ocean by the Telstar satellite.

1969 The Apollo 11 Lunar Module lands on the Moon with a crew for the first time.

1976 Vikings 1 and 2 land on Mars.

1979 The Voyager space probes photograph the planet Jupiter.

1981 First US Space Shuttle flight.

1986 The Space Shuttle *Challenger* explodes just after take-off. The orbiter is destroyed, killing all seven crew members on board.

1986 The Giotto space probe flies past Halley's Comet.

1989 The Voyager 2 space probe photographs Neptune.

1997 The Mars Pathfinder lands on Mars. A buggy called the Sojourner explores the surface and sends thousands of images back to Earth.

Loading your INTERFACT disk

INTERFACT is easy to load. But, before you begin, quickly run through the checklist on the opposite page to ensure that your computer is ready to run the program.

Your INTERFACT CD-ROM will run on both PCs with Windows and on Apple Macs. To make sure that your computer meets the system requirements, check the list below.

SYSTEM REQUIREMENTS

PC
- 486DX2/66 Mhz Processor
- Windows 3.1, 3.11, 95, 98 (or later)
- 8 Mb RAM (16 Mb recommended for Windows 95 and 24 Mb recommended for Windows 98)
- VGA colour monitor
- SoundBlaster-compatible soundcard

APPLE MACINTOSH
- 68020 processor
- system 7.0 (or later)
- 16 Mb of RAM

LOADING INSTRUCTIONS

You can run INTERFACT from the disk – you don't need to install it on your hard drive.

PC WITH WINDOWS 95 OR 98

The program should start automatically when you put the disk in the CD drive. If it does not, follow these instructions.

1 Put the disk in the CD drive
2 Open MY COMPUTER
3 Double-click on the CD drive icon
4 Double-click on the icon called SPACE TRAVEL

PC WITH WINDOWS 3.1 OR 3.11

1 Put the disk in the CD drive
2 Select RUN from the FILE menu in the PROGRAM MANAGER
3 Type D:\SPTRAVEL (Where D is the letter of your CD drive)
4 Press the RETURN key

APPLE MACINTOSH

1 Put the disk in the CD drive
2 Double click on the INTERFACT icon
3 Double click on the icon SPACE TRAVEL

CHECKLIST

● Firstly, make sure that your computer and monitor meet the system requirements as set out on page 38.

● Ensure that your computer, monitor and CD-ROM drive are all switched on and working normally.

● It is important that you do not have any other applications, such as wordprocessors, running. Before starting INTERFACT quit all other applications.

● Make sure that any screen savers have been switched off.

● If you are running INTERFACT on a PC with Windows 3.1 or 3.11, make sure that you type in the correct instructions when loading the disk, using a colon (:) not a semi-colon (;) and a back slash (\) not a forward slash (/). Also, do not use any other punctuation or put any spaces between letters.

How to use INTERFACT

INTERFACT is easy to use.
First find out how to load the program
(see page 39) then read these simple
instructions and dive in!

There are seven different features to explore. Use the controls on the right-hand side of the screen to select the one you want to play. You will see that the main area of the screen changes as you click on to different features.

For example, this is what your screen will look like when you play REPAIR SHOP, in which you have to rebuild a space probe that has been damaged. Once you've selected a feature, click on the main screen to start playing.

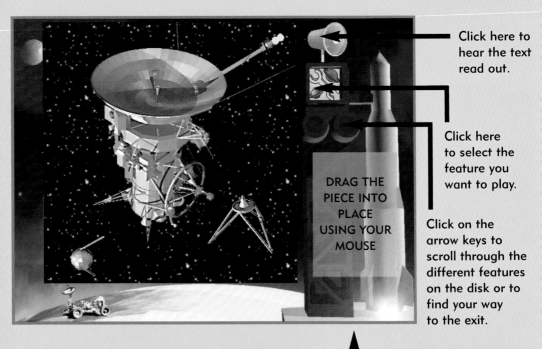

Click here to hear the text read out.

Click here to select the feature you want to play.

DRAG THE PIECE INTO PLACE USING YOUR MOUSE

Click on the arrow keys to scroll through the different features on the disk or to find your way to the exit.

This is the reading box where instructions and directions appear, explaining what to do.

DISK LINKS

When you read the book, you'll come across Disk Links. These show you where to find activities on the disk that relate to the page you are reading. Use the arrow keys to find the icon on screen that matches the one in the Disk Link.

DISK LINK
Want to find out about life on a space station? Then check out **FLOATING FREE**.

BOOKMARKS

As you play the features on the disk, you'll bump into Bookmarks. These show you where to look in the book for more information about the topic on screen. Just turn to the page of the book shown in the Bookmark.

23

ACTIVITIES

On pages 33–35, there are some fun activities for you to do. You could complete a space-themed word search or even construct your own space station!

HOT DISK TIPS

● After you have chosen the feature you want to play, remember to move the cursor from the icon to the main screen before clicking your mouse again.

● If you don't know how to use one of the on-screen controls, simply touch it with your cursor. An explanation will pop up in the reading box!

● Any words that appear on screen in a different colour and are underlined are 'hot'. This means that you can touch them with the cursor for more information or an explanation of the word.

● Keep a close eye on the cursor. When it changes from an arrow ➡ to a hand ☞ click your mouse and something will happen.

Troubleshooting

If you have a problem with your INTERFACT disk, you should find the solution here. You can also e-mail for help at helpline@two-canpublishing.com.

COMMON PROBLEMS

Cannot load disk
There is not enough space available on your hard disk. To make more space available, delete old applications and programs you don't use until 6 Mb of free space is available.

There is no sound (PCs only)
Your soundcard is not SoundBlaster compatible. To make your settings SoundBlaster compatible, see your soundcard manual for more information.

Disk will not run
There is not enough memory available. Quit all other applications and programs. If this does not work, increase your machine's RAM by adjusting the Virtual Memory (see right).

There is no sound
Your speakers or headphones are not connected to the CD-ROM drive. Ensure that your speakers or headphones are connected to the speaker outlet at the back of your computer.

Print-outs are not centred on the page or are partly cut off
Make sure that the page layout is set to 'Landscape' in the Print dialogue box.

There is no sound
Ensure that the volume control is turned up (on your external speakers and by using internal volume control).

Graphics freeze or text boxes appear blank (Windows 95 or 98 only)

Graphics card acceleration is too high. Right-click on MY COMPUTER. Click on SETTINGS (Windows 95) or PROPERTIES (Windows 98), then PERFORMANCE, then GRAPHICS. Reset the hardware acceleration slider to 'None'. Click OK. You may have to restart your computer.

Text does not fit into boxes or 'hot' words do not work

The standard fonts on your computer have been moved or deleted. You will need to re-install the standard fonts for your computer. PC users require Arial. Please see your computer manual for further information.

Your machine freezes

There is not enough memory available. Either quit other applications and programs or increase your machine's RAM by adjusting the Virtual Memory (see right).

Graphics do not load or are of poor quality

Not enough memory is available or you have the wrong display setting. Either quit other applications and programs or make sure that your monitor control is set to 256 colours (MAC) or VGA (PC).

HOW TO...

Reset screen resolution in Windows 95 or 98:

Click on START at the bottom left of your screen, then click on SETTINGS, then CONTROL PANEL, then double-click on DISPLAY. Click on the SETTINGS tab at the top. Reset the Desktop area (or Display area) to 640 x 480 pixels, then click APPLY. You may need to restart your computer after changing display settings.

Reset screen resolution in Windows 3.1 or 3.11:

In Program Manager, double-click on MAIN. Double-click on OPTIONS, then click on 'Change system settings'. Reset the screen resolution to 640 x 480, 256 colours. You will need to restart your computer after changing display settings.

Reset screen resolution for Apple Macintosh:

Click on the Apple symbol at the top left of your screen to access APPLE MENU ITEMS. Select CONTROL PANELS, then MONITORS (or MONITORS AND SOUND) then set the resolution to 640 x 480.

Adjust the Virtual Memory on your PC with Windows 95 or 98:

Open MY COMPUTER, then click on CONTROL PANEL, then SYSTEMS. Select PERFORMANCE, click on VIRTUAL MEMORY and set the preferred size to a higher value.

Adjust the Virtual Memory on Apple Macintosh:

If you have 16 Mb of RAM or more, SPACE TRAVEL will run faster. Select the SPACE TRAVEL icon and go to GET INFO in the FILE folder. Set the preferred (or current) size to a higher value.

Index

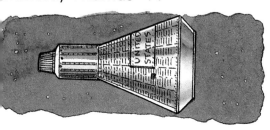

Work book

Photocopy this sheet and use it to make your own notes.

Work book